101 RIDDLES • CALCULATOR GAMES • CALCULATOR POKER
COMPLETE CALCULATOR DICTIONARY

BOGGLE

CALCULATOR WORD/NUMBER GAMES BY JAMES VINE

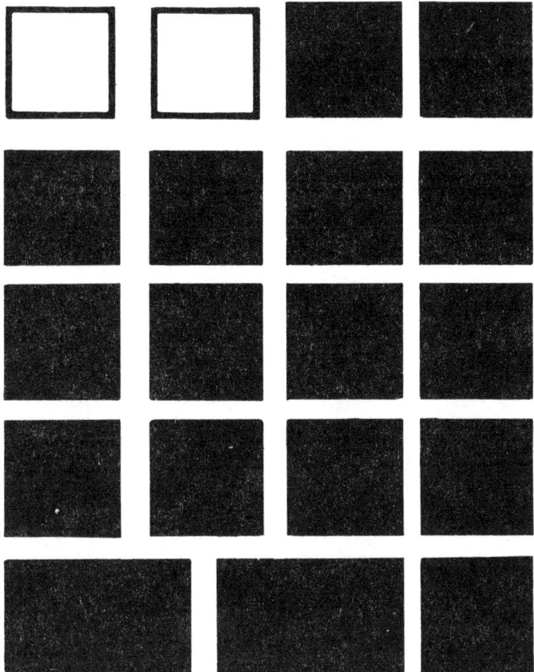

PRICE / STERN / SLOAN
Publishers, Inc., Los Angeles

INTRODUCTION

Technological advances in miniaturization and mass production in electronics have brought the hand-held electronic calculators within the financial reach of most people. In fact, in many secondary schools, it is common to allow their use in solving mathematical problems and in some, they are supplied as are textbooks.

For many years, children have told jokes which use numbers viewed upside down. (How many remember the 7734 jokes?) Therefore, it is only natural that, with the availability of small electronic calculators, there would be great interest in jokes and games which would be played with a calculator.

This book presents a collection of 101 jokes, or as I call them, Calc-o-Grams, 5 mind boggling games for 2 or more players, and a dictionary of words with their corresponding numbers.

Calc-o-Grams and the many games that can be played with your electronic calculator are fascinating ways to challenge your mind. The possibilities are almost endless.

Come BOGGLE (379908) your mind.

101
CALC-O-GRAMS
RIDDLES

(Answers are on page 63.)

Welcome to the group of thousands of calculator owners who are enjoying playing with Calc-O-Grams. Making words from numbers. Get out your calculator and, let's have some fun.

The only limitations in the ways you can make words, phrases and even sentences are your mind and the Calc-O-Gram alphabet.

In Elementary Calc-O-Grams, the alphabet is B, E, H, I, L, O, and S. These are made from the following numbers:

Letter	Number
B	8
E	3
H	4
I	1
L	7
O	0
S	5

Let's see what they look like. Put the number 5071438 in your calculator key board and turn the calculator around so that the display is upside down to you. There, we have

BEHILOS

Now let's see some other words. Put the number 77345 in your calculator key board. Now you see

SHELL

See how easy it is. Did you notice how we made the word SHELL? We
started at the <u>end</u> of the word and, with our Calc-O-Gram alphabet/number
table we found the number that is the Calc-O-Gram word, SHELL.

Here it is step-by-step. The word SHELL, backwards is LLEHS. . Using our
table we see that this is

L	7
L	7
E	3
H	4
S	5

or 77345

Making words from numbers is fun, but in Calc-O-Grams you need a story
to go with it. Here is an example:

1. What do 55 snakes in a 14 foot square pit do?

Story	Keyboard and Function	Display Shows
What do 55 snakes in a 14 foot square pit do?	Enter 55 Enter 14	55 5514
Turn your calculator around and you see <u>HISS</u>		

We used the word Function with Key Board. In Elementary Calc-O-Grams we
have 5 Functions.

                         ENTER
                         ADD
                         SUBTRACT
                         MULTIPLY
                         DIVIDE

You've just learned how to use the first one, ENTER. We'll get to the
others later after you've had some practice with ENTER.

Mathematicians like to use symbols, probably to save writing, but it also
makes the rules clear. So that's what we're doing now.

The symbol for the function is the first letter of the function's name.

Symbol	Function
E	ENTER
A	ADD
S	SUBTRACT
M	MULTIPLY
D	DIVIDE

ENTER is not a mathematical function and there is not a single key
on your calculator for it, as there are for the other four. Enter means
to put the numbers following into the number key board in the sequence
that the numbers are listed.

<table>
<thead>
<tr><th>Story</th><th>Keyboard and Function</th><th>Display Shows</th></tr>
</thead>
</table>

Here are two more samples of ENTER.

2. If (E) 3 men walked	E 3	3
(E) 510 miles in	E 510	3510
(E) 8 weeks, without seeing anyone,	E 8	35108
name a city in Idaho that they		
could be near.		BOISE

3. Name a state that has (E) .140	E .140	0.140
automobiles for every mile of		
roads.		OHIO

Notice we used a decimal point before the number. This is because most calculators won't show a 0 in front of a number. We make it appear by starting with a decimal point and the calculator puts a zero in front of the decimal.

Remember, if the last letter of the word you want to make is a "O" then use a decimal point instead of a 0 (zero).

Let's try 2 more examples before we learn ADD and SUBTRACT.

4. Why do (E) 535 people like	E 535	535
to work for (E) 50 people for	E 50	53550
(E) 8 hours? Because they have	E 8	535508
understanding _________.		BOSSES

5. When you want butter, but
    can only spend (E) $.370,
    what do you get?   E .370   0.370
                                                            OLEO

More examples of ENTER are shown later. Before we start with ADD and
SUBTRACT, notice the words we used in the stories for ENTER. As in all
mathematical type word puzzles, the words used, or how the numbers are
stated, tell you which Function to use.

For ENTER we use

1. A statement of numbers.
2. Statements of numbers separated by a comma, ⎯/ (e.g.) 46, 22, 36 ⎯/
3. The words IN, WITH, ABOUT.

At this point, we will introduce a convention for showing how the Calc-O-
Grams are "played" on your calculator. This will eliminate the problems
with differences in how calculators operate and confusion that could exist
about the exact steps to be taken in forming the words.

The symbol E will no longer be used. The numbers shown are to be treated
with the mathematical function (A, S, M, or D) as indicated. A sequence
of numbers is entered until another mathematical function appears. All
answers shown under "Display Shows" are as though the = key was pressed
after the sequence of numbers had been entered with the proper mathematical
function.

Now let's try ADD (A) and SUBTRACT (S). Notice the words used to tell
what Function to use.

Story	Keyboard and Function	Display Shows
6. In the summer there was a young man who fell madly in love with a petite young lady who measured 31, 22, 31.	312231	312231
He always brought candy when he saw her, <u>and</u> (A) in only 3 months he traveled 250 miles to see her	A 3250	315481
<u>and</u> (A) found that she had gained 57 pounds. He thought she so looked like	A 57	315538
a little cow, he called her his ____.		<u>BESSIE</u>

Let's go over this example. The numbers we added were

$$312231 + 3250 + 57 = 315538$$

The word we used to say ADD was <u>and</u>. Whenever <u>and</u> appears, ADD the
number following. If no mathematical function is indicated, simply
ENTER a number.

Story	Keyboard and Function	Display Shows
7. Working for GM 50 men, make $8050 per year	508050	508050
(S) 5 of them <u>lost</u> their jobs, used up their unemployment. What do you think they	S 5	508045
are now?		<u>5 HOBOS</u>

Notice, we used a number to represent a number. There are 6 numbers that are numbers in Calc-O-Grams. These are

Number	Calc-O-Gram Number
0	0
1	1
2	2
5	5
6	9
9	6

Here are the words in the stories that are used for ADD or SUBTRACT

ADD

1. And
2. With
3. Or any word or phrase which says there is more now than before.

SUBTRACT

1. Lose, lost, etc.
2. Went away
3. Or any word or phrase which says there is less now than before.

Story	Keyboard and Function	Display Shows
8. 1 man wrote a bad story about 7 politicians	17	17
and he sold the story to (A) 800 newspapers	A 800	817
and it was read by (A) 73000 people. What do you think	A 73000	73817
the politicians sued him for?		LIBEL

Story	Keyboard and Function	Display Shows
9. 39 immigrants	39	39
<u>and</u>		
<u>(A)</u> 4 of their relatives are all from Canada. What's the word they all use to end a sentence or ask a question?	A 4	43
		<u>EH</u>
10. Walter went fishing. He caught 15 fish	15	15
<u>and</u> (A) 2 old tires	A 2	17
<u>and</u> he said he caught		
<u>(A)</u> 150 bass	A 150	167
<u>and</u> (A) 95 halibut	A 95	262
<u>and</u> (A) 55 perch.	A 55	317
His story was a _____ .		<u>LIE</u>

Now let's try MULTIPLY, whose symbol is M.

Story	Keyboard and Function	Display Shows
11. 37 men were listening to a salesman tell the wonders of his gold mine stock.	37	37
<u>Each</u> (M) bought 15 shares	M 15	555
at (M) $139. <u>each</u>	M 139	77145
Wouldn't you expect that there was a _____ among them?		<u>SHILL</u>

Story	Keyboard and Function	Display Shows

The words meaning MULTIPLY are

    1. Each.
    2. At # each, or of # each.
    3. Times.
    4. Or any word meaning multiply.

Here are four more examples of MULTIPLY

Story	Keyboard and Function	Display Shows
12. What would some people call George if he stole 5 welfare checks <u>each</u> for (M) $161.?	5 M 161	5 805 <u>SOB</u>
13. 2 Christians walked 101 miles <u>each</u> for (M) 9 days. <u>each</u> carrying (M) 2 large packages to buy this book. What is it?	2101 M 9 M 2	2101 18909 37818 <u>BIBLE</u>
14. 2 sailors, saw 53 girls in 8.67 minutes. At <u>each</u> they waved (M) 2 times. They said a lot of _____.	2538.67 M 2	2538.67 5077.34 <u>HELLOS</u>

Story	Keyboard and Function	Display Shows
15. 601 children, learning to swim, practiced <u>each</u> (M) 3 minutes doing the frog kick. They <u>each</u> did (M) 25 kicks per minute. They sure did ______ the water.	601 M 3 M 25	601 1803 45075 <u>SLOSH</u>

Now for DIVIDE, whose symbol is D.

The words meaning DIVIDE are:

    1.   Share.
    2.   By.
    3.   Divide.
    4.   Or any word meaning among or divide.

Story	Keyboard and Function	Display Shows
16. If 501 GI's got K.P. and had to <u>share</u> peeling (D) 12500 onions. They each would probably ______.	501 D 12500	501 0.04008 <u>BOOHOO</u>
17. 9 women were frightened <u>by</u> (D) 200 ducks. To scare them away they yelled _____.	9 D 200	9 0.045 <u>SHOO</u>
18. The Barstow Music Society needed to have an opening number for their spring concert. They couldn't decide so they selected 1 tenor, 4 baritones, 1 bass <u>out of</u> (D) 200 singers to try out for the ______.	141 D 200	141 0.705 <u>SOLO</u>

	Story	Keyboard and Function	Display Shows

19. 2 womens' clubs went to a concert
and <u>shared</u> (D) 25 minutes listening
to an off tune tenor. When he
finished, there was a loud _____.

	2	2
	D 25	0.08
		<u>BOO</u>

20. Santa Claus had 40 hours
until Christmas Eve. He was
worried until he decided to
<u>share</u> the work among his
(D) 99 elves. Then he went
around saying __ __ __ __.

	40	40
	D 99	0.4040404
		HOHOHOHO

You have seen five examples of each function. Now let's put some together
for multiple function Calc-O-Grams.

Here first are examples of MULTIPLY and ADD or SUBTRACT.

21. 3 men bought 1, $400 horse,
<u>and</u> they bought a $167 saddle
The horse kicked them <u>each</u>
(M) 12 times. To stop him
they had to _______ him.

	31400	31400
	A 167	31567
	M 12	378804
		<u>HOBBLE</u>

Story	Keyboard and Function	Display Shows
22. 10 neighbors, lived on 1 floor of an apartment house. They had a noisy musician on their floor. They plotted <u>each</u> (M) 61 minutes <u>each</u> paid (M) $5 plus (A) $.14 to get rid of ___ ___.	101 M 61 M 5 A .14	101 6161 30805 30805.14 <u>HIS OBOE</u>
23. John had 77 rings, 18 seconds apart done on a church bell (M) 10001 <u>times</u>, <u>and</u> he paid only (A) $20. It was a cheap ___ ___.	7718 M 10001 A 20	7718 77187718 77187738 <u>BELL BILL</u>
24. In an evening, 70 girls <u>each</u> earned (M) $501 <u>less</u> (S) $63 cab fare. One would suspect that they are ___.	70 M 501 S 63	70 35070 35007 <u>LOOSE</u>
25. If you take $550 to the horse races <u>each</u> of (M) 10 days <u>and</u> have only (A) $7 at the end. You have sustained quite a ___.	550 M 10 A 7	550 5500 5507 <u>LOSS</u>

Story	Keyboard and Function	Display Shows

Let's have some fun with DIVIDE and ADD or SUBTRACT Calc-O-Grams.

Story	Keyboard and Function	Display Shows
26. At the Miss Yuma Beauty Contest they were looking for a girl with a figure of 36, 25, 36 <u>and</u> the (A) 61 judges selected her <u>out</u> <u>of</u> (D) 71 contestants. Her name was ______.	362536 A 61 D 71	362536 362597 5107 <u>LOIS</u>
27. 16 sheiks had 59 square miles to <u>share</u> with their (D) 30,000 tribesmen <u>and</u> they found (A) 695 businessmen <u>and</u> (A) 15 bankers to form ____ ____.	1659 D 30000 A 695 A 15	1659 0.0553 695.0553 710.0553 <u>ESSO</u> <u>OIL</u>
28. 15 politicians, 46 lobbyists, 9 bankers met to see how they could <u>share</u> (D) 20,000 acres of off-shore leases <u>and</u> they talked with (A) 40 senators <u>and</u> (A) 150 civil servants <u>and</u> (A) 47 influential stock brokers <u>and</u> (A) 473 investors. Out came a contract for ______ ___.	15469 D 20000 A 40 A 150 A 47 A 473	15469 0.77345 40.77345 190.77345 237.77345 710.77345 <u>SHELL</u> <u>OIL</u>

Story	Keyboard and Function	Display Shows
29. What do you get stung by if you find a ball in a tree up 6 feet, 7 inches long, 6 inches across after you break it into (D) 2 parts? A ___.	676 D 2	676 338 BEE
30. If 101 women at a club luncheon had to share (D) 200 pieces of chicken. They would say that the lunch was ___ ___.	101 D 200	101 0.505 SO SO

Now the most complex Calc-O-Grams using MULTIPLY, DIVIDE and ADD or SUBTRACT and we make phrases and sentences. The next 4 Calc-O-Grams make the sentence HE LOSES HIS 92 LBS.

Story	Keyboard and Function	Display Shows
31. Here's what happened to a young Englishmen who decided he would gamble all of his pay check. He took his 92 pounds, walked from his office back to (S) 58 Farthing Place. ___	92 S 58	92 34 HE

	Story	Keyboard and Function	Display Shows
32.	He decided to <u>divide</u> his money in (D) 2 parts and to <u>multiply</u> his money by betting on horse (M) 3 race 1, jockey 4, gate 8 His horse came in <u>behind</u> the winner by (S) 9 lengths.	Leave the number 34 in. D 2 M 3148 S 9	17 53516 53507 <u>LOSES</u>
33.	Worse things are yet to come. He discovered that he had <u>lost</u> (S) 51 pounds. So he decided to go for broke and <u>divided</u> his bet with horse (D) <u>10</u>, horse 4. Now we have	Leave the number 53507 in. S 51 D 104	53456 514 <u>HIS</u>
34.	He went to <u>multiply</u> the last of his money by betting at bookie (M) 1, window 15 His horses were in a photo finish for <u>last</u>, coming in (S) 38 lengths, 4 feet <u>behind</u> the winner.	Leave the number 514 in. M 115 S 384	59110 58726 <u>92 LBS</u>

Now we've made HE
        LOSES
        HIS
        92 LBS.

<table>
<tr><td align="center">Story</td><td align="center">Keyboard and
Function</td><td align="center">Display
Shows</td></tr>
</table>

Here's another 4 part Calc-O-Gram sentence.

This one says HIS ISLE IS BLISS.

Story	Keyboard and Function	Display Shows
35. A young man took his last 2 pay checks of <u>each</u> (M) $257 and bought a small sailboat.	2 M 257	2 514 <u>HIS</u>
36. He headed the boat west <u>and</u> sailed (A) 3237 miles to a beautiful tropical ______.	Leave the number 514 in. A 3237	3751 <u>ISLE</u>
37. He went ashore <u>and</u> met (A) 23 lovely native girls. He <u>divided</u> the next (D) 74 hours into just having fun.	Leave the number 3751 in. A 23 D 74	3774 51 <u>IS</u>
38. During the following days his contentment <u>multiplied</u> (M) 1081% <u>and</u> within (A) 47 hours decided to never leave.	Leave the number 51 in. M 1081 A 47	55131 55178 <u>BLISS</u>

<table>
<tr><td>Story</td><td>Keyboard and Function</td><td>Display Shows</td></tr>
</table>

Here are 63 more Calc-O-Grams for fun and learning. Try them all on your calculator. Answers are on page 63.

	Story	Keyboard and Function	Display Shows
39.	16 people with (A) 42 pounds of baggage went on a bargain trip to Hawaii for (A) \$79. They only got one ______.	16 A 42 A 79	16 58 137 ______
40.	If the 16 people had paid each (M) \$321.07 for first class fares, they would have gotten __________.	16 M 321.07	16 5137.12 __ ______
41.	2 men, 5 boys, 7 girls were each locked in a cookie factory over a (M) 3 day holiday. They each ate (M) 46 packages of cookies. The factory said it had lost (S) \$86. in cookie sales because they had eaten so much. If they did that many weekends, they would become fat or __________.	257 M 3 M 46 S 86	257 771 35466 35380 ______
42.	There is a woman who has 5 daughters answering her phone number of 7735345 The police wonder what she does. We know. ______ ______.	57735345	57735345 ______ ______

Story	Keyboard and Function	Display Shows

43. The police <u>left</u> (S) 5 men on the
phone 715 hours, listening to
8000 calls
They found that she only sells

_______ .

	Keyboard and Function	Display Shows
	Leave the 57735345 in.	
	S 57158000	577345
		______

44. There was a young lady who weighed
100 pounds. She entered an essay contest
for a candy company and won <u>each</u> day
(M) 6 candy bars
<u>each</u> for (M) 263 days. She got fat
and went on a diet and <u>lost</u> (S) 31
pounds in <u>each</u> of 2 months. Her
boyfriend said she still looked like
a little cow, so he called her his

_______ .

	Keyboard and Function	Display Shows
	100	100
	M 6	600
	M 263	157800
	S 31	157769
	M 2	315538
		______

45. She got mad <u>and</u> threw 2 pans, 1
skillet, 80 dishes at him
because she was stubborn. Her
name was _______ .

	Keyboard and Function	Display Shows
	Leave 315538 in.	
	A 2180	317718
		______

46. There was a sultan who had 3 eunuchs
56 wives, which he visited
(M) 20 <u>times</u> a year. This visit
he found <u>missing</u> (S) 1 eunuch, 2 wives.
So he said I will _______ him.

	Keyboard and Function	Display Shows
	356	356
	M 20	7120
	S 12	7108
		______

Story	Keyboard and Function	Display Shows
	Leave 7108 in.	
47. He added (A) 2 more guards to the search party searching for the eunuch, <u>divided</u> his wives and children	A 2	7110
<u>into</u> (D) 10 groups	D 10	711
<u>leaving</u> (S) 1 group to get	S 1	710
the ____.		____
48. 2 priests, 76 nuns, 89 laymen	27689	27689
<u>each</u> saw the old man (M) 2 times	M 2	55378
so they could _______ him.		____
49. 2 bishops, 23 priests, 7 nuns	2237	2237
<u>each</u> said (M) 19 prayers	M 19	42503
<u>each</u> hour for (M) 18 hours	M 18	765054
<u>each</u> for (M) 7 days, so that	M 7	5355378
the archbishop could		
say he _______ the chapel.		____
50. 1 man told 377 workers to	1377	1377
<u>each</u> spend (M) 4 hours	M 4	5508
picking apples. He was		
their ____.		____

Story	Keyboard and Function	Display Shows
51. 80 school children **each** had (M) 101 cans of paint. They **lost** (S) 2 of them when they **fell** down the stairs. At the bottom was a big ____ .	80 M 101 S 2	80 8080 8078
52. George was sent to buy some shirts for his rally team. He spent $3.87 **each** for (M) 41 shirts at **each** of (M) 2 stores. He plans to **add** (A) $.04 more for the shirts and say it cost him that much. If he does, he will ____ the truth.	3.87 M 41 M 2 A .04	3.87 158.67 317.34 317.38
53. 1 boy, 3 girls **each** hit (M) 2 hives **each** (M) 13 times until they were stung by a ____ .	13 M 2 M 13	13 26 338

Story	Keyboard and Function	Display Shows
54. 8 girls each wanted the phone number to apartment (M) 101 because that's where bachelor ____ lived.	8 M 101	8 808 ———
55. 3 girls, 7 boys watched the water in the bay each for (M) 24 hours (S) 5 left because they were tired of watching the tide ____.	37 M 24 S 5	37 888 883 ———
56. What do you get when you dig something 29 inches by (M) 16 inches less (S) 1 inch by (M) 8 inches? ____	29 M 16 S 1 M 8	29 464 463 3704 ———
57. A young man spent 9 days each of (M) 20 months only to lose (S) 7 pounds while inventing a Cotton Gin. His name was ____ Whitney.	9 M 20 S 7	9 180 173 ———

Story	Keyboard and Function	Display Shows
58. 9 men <u>each</u> spent (M) 397 hours trying to learn to fly. After many efforts they decide to try something ____.	9 M 397	9 3573 ————
59. 4 sailors, 2 marines, 1 airman <u>each</u> spent (M) 75 hours painting her house. They <u>lost</u> (S) 2 brushes, all for the love of ____.	421 M 75 S 2	421 31575 31573 ————
60. 2 cousins <u>each</u> (M) 17 years old, when explaining a broken vase will say "___ did it".	2 M 17	2 34 ——
61. Charlie stole $611 from <u>each</u> of (M) 6 ladies <u>plus</u> (A) 1 automobile. He had so much fun he did it (M) 2 <u>times</u>. What a ____!	611 M 6 A 1 M 2	611 3666 3667 7334 ————
62. 3 blonds, 6 brunetts, 7 redheads <u>each</u> walked (M) 200 steps in front of <u>each</u> of (M) 8 judges at the beauty contest. The judges <u>added</u> up all their assets and finally picked (A) 1 blonde who only weighed ———— ————	367 M 200 M 8 A 1	367 73400 587200 587201 ———— ————

Story	Keyboard and Function	Display Shows
63. When you are sailing, if you throw 14 pieces of paper up in the air <u>each</u> (M) 24 times <u>and</u> try to catch them (A) 1 at a time, they will always fall on the _____ side.	14 M 24 A 1	14 336 337 ___
64. 1 man, 1 woman set sail <u>each</u> traveling (M) 341 miles until they met on their secret _____.	11 M 341	11 3751 ___
65. 7 men went to Las Vegas <u>Each</u> took (M) $791. their luck was bad and they came back with much ____ than they started.	7 M 791	7 5537 ___
66. Tom tells everyone he meets, that he bought 2 Cadillacs each for (M) $2658.67. They're either old or ____ _____.	2 M 2658.67	2 5317.34 ___ ___
67. 160 men <u>each</u> tried (M) 173 hours to find the secrets of the magic wand. (S) 1 man <u>left</u> and the rest <u>each</u> tried (M) 20 times again. The secrets really ____ them.	160 M 173 S 1 M 20	160 27680 27679 553580 ___

Story	Keyboard and Function	Display Shows
68. A clarinet player didn't like the sound of his instrument. He worked 11 hours to make a new instrument each week for (M) 7 days for each of (M) 40 weeks. Then he liked his new ______.	11 M 7 M 40	11 77 3080
69. 5 sailors on leave each spent (M) 67 minutes waiting to _____ a certain girl.	5 M 67	5 335
70. They each saw (M) 948 girls and left (S) 43 gifts before they found ______.	Leave 335 in. M 948 S 43	317580 317537
71. 3 men each (M) 23 years old each spent (M) 5 hours looking for a certain ______	3 M 23 M 5	3 69 345
72. 5 men went sailing each (M) 101 miles from shore. Their boat began to take on water, so they sent out an ______.	5 M 101	5 505

	Story	Keyboard and Function	Display Shows
73.	A farmer bought 5 acres of land at (M) $1421 <u>each</u> because it had fertile ______.	5 M 1421	5 7105 ————
74.	8 men <u>each</u> worked (M) 40 hours <u>each</u> made (M) $2417. When they got their checks and saw how much in taxes they had paid, they said ___ ____.	8 M 40 M 2417	8 320 773440 ___ ____
75.	Porgy went shopping one day with 5 dimes 5 nickels 38 pennies because that's all he had to buy _____ a present.	5 5 38	5 55 5538 ————
76.	3 Philippino men 5 women, 7 children 1 goat <u>each</u> walked (M) 71 miles <u>each</u> for (M) 2 days looking to go to war. They finally gave up because they only had __ ____.	3 5 7 1 M 71 M 2	3 35 357 3571 253541 507082 __ ____

Story	Keyboard and Function	Display Shows
77. If Kissinger, after meeting with 4 heads of state, 21 minor dignitaries, <u>each</u> for (M) 41 hours on <u>each</u> of (M) 3 trips, can't solve the old problem, maybe he'd better call on the Moslem Satan ______.	4 21 M 41 M 3	4 421 17261 51783 ______
78. What's 7 feet long, 3 inches in diameter, eats 3 pounds of fish a day and lives in the ocean?	7 3 3	7 73 733 ___
79. Who benefits when you take 710 sheiks 77 businessmen, give them 3 years to develop 45 square miles of sand?	710 77 3 45	710 71077 7107.73 71077345
80. John paid $53. to play in a bowling tournament. He bowled 5, 0, 7, 3 his first 4 frames. At the end of the tournament the announcer said that unfortunately ___ ______ his money	53 50734	53 5350734 ___ ______
81. George walked 73 miles in 34.5 hours carrying a 14 pound bag. He complained that ___ _____ hurt.	73 34.5 14	73 7334.5 7334.514 ___ _____

Story	Keyboard and Function	Display Shows
82. Mrs. Johnson found her husband's		
7 credit cards	7	7
went to 7 stores	7	77
bought $18.51 of merchandise	18.51	7718.51
in 4 stores. Mr. Johnson	4	7718.514
got a surprise at the end		
of the month, _____ _____.		_____ _____
83. 3 sailors spent	3	3
75 days	75	375
traveling 1514 miles to search	1514	3751514
for something that was used by		
Captain Kidd. It was _____ _____.		_____ _____
84. 55 girls started on a hike	55	55
37 miles long,	37	5537
51 finish. The number	51	553751
finishing ___ _____ than the		_____ _____
number that started.		
85. 11 men, 80 years old, said 1 prayer	11801	11801
each (M) 157 times,	M 157	1852757
each of (M) 2 hours. The devil	M 2	3705514
appeared, offered to grant them		
their wish if they each would sell		
_____ _____.		_____ _____

Story	Keyboard and Function	Display Shows
86. Bobby Riggs bet 5 men $80. that in 7.3 hours he could out play 4 in tennis. Bobby lasted and won, because in his tennis game ____ ____.	5 80 7.3 4	5 580 5807.3 5807.34 __ __
87. There was an old lady who lived in a shoe. She had 30 children owed $4551 on her mortgage the banker gave her 4 days to pay up or her home would become ____ ____.	30 4551 4	30 304551 3045514 __ __
88. 8 foremen <u>and</u> (A) 495 supervisors <u>and</u> (A) 5005 machinists all work for one man, their ____.	8 A 495 A 5005	8 503 5508 ____
89. 49,000 reindeer go north every year <u>and</u> travel (A) 1500 miles <u>and</u> 307 are killed by this type of wolf. What is it?	49000 A 1500 A 307	49000 50500 50807 ____
90. After an argument with your neighbor, if you piled 6800 tons of dirt <u>and</u> (A) 914 tons of rock in his back yard, you'd be making a mountain out of a mole ____.	6800 A 914	6800 7714 ____

By now you should not need any help with key board or function use.
The balance of the Calc-O-Grams have the function words underlined and
at the end, the mathematical formula and words are shown.

91. What insect travels 53 miles _and_ works 6 hours a day _and_ travels
     in groups of up to 250 _and_ lives 29 days?

                 53 + 6 + 250 + 29 = 338   ____

92. 5 girls had $800 _and_ they bought 72 quarts of ice cream _and_ 4 quarts
     of chocolate topping. They ate it all and each gained __ ____.

                 5800 + 72 + 4 = 5876   _ ____

93. 4 Cowboys in town, for the weekend _each_ shot 67 windows in _each_
     of 3 city blocks. They sure did raise ______.

                 4 X 67 X 3 = 804   ____

94. Out of every 8 women, 1 will say she does, 7 will say they don't
     support women's ___.

                 817   ____

95. A publisher every month sells 37,500 copies of a book _and_ some
     copies sell for $250, _and_ some $65 _and_ some $3. It is the world's
     best selling book. What is it?

                 37500 + 250 + 65 + 3 = 37818   ______

We will end our sample Calc-O-Grams with a group which form a sentence
dedicated to the ecologist. The sentence is SEE SHELLS OIL BLOB, or
if you use the substitutions, SEE ESSOS OIL BLOB. Do not clear your
calculator betwen each Calc-O-Gram in the sentence.

96.   The race was on for off-shore oil. 6 oil companies, 7 cat wildcatters
      each built 5 platforms.

              67 X 5 = 335   ____

97.   Each started drilling in 1723 feet of water and at 140 feet an
      Oil Company hit rock.

          335 X 1723 + 140 = 577345   ______

98.   They decided to remove the drill by 115 feet and divide
      the rock with 813 sticks of dynamite.

          (577345 - 115) ÷ 813 = 710   ____

99.   It worked. They got enough oil to fill each of 11 tankers and
      have 268 more.

          710 X 11 + 268 = 8078   ____

   If you prefer to say ESSO, then substitute 100 and 101 for #97 and #98.

100.   Each started drilling in 150 fathoms of water and at 303 fathoms
        an Oil Company hit rock.

          335 X 150 + 303 = 50553   ______

101.   They decided to remove the drill by 143 fathoms and divide the rock
       with 71 pounds of TNT.

          (50553 - 143) ÷ 71 = 710

Calc-O-Gram Games not only are fun, but they are mental challenges. The more you play them, the greater your skill will become and you will find that you enjoy having more difficult games to play.

These games have an element of chance, as drawing certain numbers, but the fun and skill is that two players with the same numbers will get different scores. Skill makes the difference.

For the first time, here are games that truly combine fun, skill and chance. The more experienced player has an advantage only at first. Other players quickly gain experience and then it is a game of skill and mental challenge.

The objective of all Calc-O-Gram Games is to make words with numbers in the calculator display, when the numbers are viewed upside down.

Here are five Calc-O-Gram Games, starting with the simplest followed by progressively more difficult challenges. Play them and have fun.

All acceptable words (and some abbreviations) will be found in the 9 letter Alphabet Dictionary. The Dictionary, which is sequenced alphabetically and numerically should be used during play by all players.

More than one word can be made at a time and the group of words does not have to make sense. (As I BOB SHOE is acceptable).

To play these Calc-O-Gram Games, you will need.

1. An 8 digit electronic calculator, for each player, with add, subtract, multiply and floating decimal point. A memory is not required.

2. A deck of playing cards with the face cards and jokers removed.

3. A three minute egg timer.

4. Pencils and paper and,

5. Poker chips.

Get your calculators, pencils, paper, cards and timer out, and here we go.

<u>ADD-ON</u> <u>CALC-O-GRAMS</u>

Number of players: 2 or 3 as single players
                      4 or 6 as partners

The Game ADD-ON

   Each player is dealt a hand of 5 cards, where the cards have a
   value of from 1 through 10. The players select one card and
   "play" the number in the calculator. The number may be played
   in any position of the calculator's display by adding as many
   zeros to the right as needed. (i.e. 5 may be entered as 50,
   500, 5000, 50000, etc.) If <u>all</u> numbers in the calculator form
   a word or words when viewed upside down, the player or team
   scores 1 point. The first player or team to reach 21 points
   with at least 2 points more than any other player, wins.

Equipment required.

   1.   Two, 8 digit electronic calculators. One calculator is
        called the Game Calculator, the other the Play Calculator.
        (Each player may wish to have his own).

   2.   Pencil and paper to keep score.

   3.   A one minute timer. (Optional)

   4.   A deck of playing cards with all face cards and jokers
        removed. When there are 4 or more players, 2 decks
        are put together and used.

   5.   The 9 Letter Alphabet Dictionary in alphabetic and
        numerical sequence.

Dealing the cards

   The cards are shuffled and spread face down on the table. All
   players select a card and the player with the highest card is
   the dealer for the game. (Suits from highest to lowest value
   are Spades, Hearts, Diamonds then Clubs).

   The dealer shuffles the cards and the player to the right cuts
   the cards.

   The cards are used to select a starting number for the Game
   Calculator. The top card is dealt face up and its value is

entered into the Game Calculator. The second is dealt face up
and its value is entered. This is repeated until 4 numbers
have been selected and a 4 digit number is in the Game Calculator.
If a 10 is dealt, it is ignored and the value of the next non-
10 card is used.

The cards used to select the starting number are placed in a
discard file.

Each player is dealt a hand of 5 cards, one card at a time,
beginning with the player to the left of the dealer and continuing
clockwise. As many hands are dealt and played as are necessary
for one player or team to win.

When all cards have been dealt, the discard file is shuffled and
cut as before and, the deal continues.

The dealer is selected for the entire game, i.e. until a player
or team wins .

Playing-Singles

Play begins with the player to the left of the dealer and
continues clockwise. Each player attempts to make a "word"
in his turn. Play continues until all cards have been used.

Playing-Partners

The partners are seated side-by-side to allow them to confer
during play.

Play begins with the partners to the left of the dealer, and
his partner. The partners select a card from either of their
hands and make their play. The partners to their left then
select one card from either of their hands, etc.

When one partner of a team has used all of his cards, play
must be made from the other partner's hand until all cards
are used.

Playing

Play continues with as many hands as needed until one player
or team has 21 points with at least 2 points more than any
other player or partners.

The number is <u>left</u> in the Game Calculator between hands and is only reset at the <u>end</u> of a game.

The Play Calculator may be used to test various alternative plays. If each player, or set of partners, does not have a calculator, then the Play Calculator is passed on with the play.

<u>Sample Hands and Plays of ADD-ON</u>

The starting number selected is   8916

Hands are:

Player A	Ace	Spades	Player B	2 Clubs
	5	Hearts		3 Spades
	6	Spades		4 Clubs
	7	Hearts		5 Diamonds
	10	Spades		9 Hearts

Here is a sample play of the first two hands of the game. Player A begins.

Player A Card	Value	Player B Card	Value	Game Calculator 8916 start	Word (s)
6 Spades	-600			8316	
		9 Hearts	-9	8307	
Ace Spades	-1000			7307	
		4 Clubs	+400	7707	LOLL
5 Hearts	+50,000			57707	LOLLS
		5 Diamonds	-50,000	7707	LOLL
7 Hearts	+7			7714	HILL
		2 Clubs	+20	7734	HELL
10 Spades	+10,000			17734	HELL I 2 words
		3 Spades	+300,000	317734	HELL IE 2 words

Player A has 3 points
Player B has 4 points

The next hands are

	Player A			Player B	
Player A	4 Diamonds		Player B	Ace Hearts	
	5 Spades			2 Spades	
	6 Diamonds			4 Clubs	
	7 Clubs			8 Clubs	
	9 Spades			10 Hearts	

*Note the 10 can be considered to be the same as an Ace except that it must be played as 10, 100, 1,000, etc. and not 1.

Play continues

Player A		Player B		Game Calculator	Word
Card	Value	Card	Value	317734	HELL IE
4 Hearts	-400			317334	HEEL IE
		2 Spades	+200	317534	HES LIE
7 Clubs	+7			317541	IHS LIE
		10 Hearts	-10,000	307541	
5 Spades	+500,000			807541	IHS LOB
		8 Clubs	+8000	815541	IHS SIB
6 Diamonds	-6000			809541	IHS GOB
		4 Clubs	-400	809141	I HI GOB
9 Spades	-90			809051	I SO GOB
		Ace Hearts	-1000	808051	I SO BOB

After 2 hands

Player A has 8 points
Player B has 8 points

Strategy of Playing

The numbers displayed in the Game Calculator may be considered to be any word or words found in the 9 Letter Alphabet Dictionary and their plural or possessive form (as BOB'S). The ending number in the example was:

808051 which may be	I SO BOB	808 05 1
	IS OB OB	80 80 51
	I SOB OB	80 805 1

and all represent words.

Therefore, when playing on a multiple word number, you can consider
playing on <u>each</u> word in the number. In addition, when part of the
number is a word and the balance is not, you may consider playing
on that part which is not a word, (e.g. the above number 808 051
(ISO BOB) 808 is a word and 051 is not). Play may be tried on 051.

Here is a table which shows possible plays with cards from Ace
through 10 on each of the words contained in 808051, plus the non
word 051 where the balance of 808051 is a word. In this table we
show plays for addition and subtraction and the number of zeros to
be added to the number played.

Number	1		05		51		051		80		805		808	
How card played	+	-	+	-	+	-	+	-	+	-	+	-	+	-
Ace	None			-1/04		-1/50	None		+1/81		+10/815	-1/804	+1/809	-1/807
2	None		None			-20/31	+20/071		None		+2/807		None	
3	None		None		+30/81		None			-30/50	+3/808			-3/805
4	None		+4/09	-4/01	None		None			-40/40	+4/809		-400/408	-4/804
5	None		None		None		None		+5/85		None		None	
6	None		None		None		None			-60/20	None		None	
7	None		None		None		None			-7/73	None		+7/815	-70/738
8	None		None			-8/43	None		None			-800/005	None	
9	None		+9/14		None		None		None		None		+9/817	
10	None			-1/04		-1/50	None		+1/81		+10/815	-1/804	+1/809	-1/807

Note: 805 for Ace and 10, -100 gives 705
        808 for Ace and 10, + 10 gives 818, -100 gives 708, + 100 gives 908.

In addition there are alternatives where an addition or subtraction
carries over to an adjacent word.

Example:
```
       80 805 1
     +    700 0
       81 505 1        1 SOS IB or 1 SO SIB
```

The above table shows us that regardless of what cards you held, you
could make a play to win a point.

The alternatives are unlimited!

General Rules

Time for play.

> The players agree on a rule for the time that a player has to
> take his turn. It is recommended that when players are learning
> ADD-ON, they be given 2 minutes to make their play. After
> experience is gained, the time should be set at 45 seconds to
> 1 minute.

Making the play on the Game Calculator.

> Once the number has been added to, or subtracted from the number
> in the Game Calculator, the player's turn is over, even though
> he made a mistake. If the Game Calculator number was accidentally
> cleared, multiplied or divided, the Game Calculator is reset to
> the number that was in it, prior to the error.

The number in the Game Calculator.

> The decimal point for the number is always to the extreme right
> of the number.

> Those number words starting with 0 (zero) cannot be used in
> the extreme left side of the number.

> The only digits considered in the number are those that are
> displayed on the Game Calculator.

<u>CREATE</u> <u>Calc-O-Grams</u>

Number of players: 2 or more

The Game CREATE

Each player is dealt a hand of 5 cards, having the value of
from 1 through 10. The players, in a fixed time, CREATE a
number word by adding, subtracting, multiplying and/or dividing
the numbers in their hand by each other. The players use as
many of their 5 cards as they can. The numbers may be used
after adding as many zeros to their right as desired. (5 may
be used as 5, 50, 500, 5000, etc.)

Players are scored for each hand by the sum of the value of
the digits of the number word formed (i.e. 77345, scores
7 + 7 + 3 + 4 + 5 = 26).

Bonus points are given for

        Using all 5 cards   10 points
        Using Multiply   10 points
        Using Divide   10 points

The deal moves to the left and another 5 card hand is dealt, etc.

The first player to make 200 points wins.

Equipment required

1.   An 8 digit electronic calculator for each player.

2.   Pencil and paper that is lined and ruled. Graph or
     accountant's paper is best. See examples for the use
     of the practice paper.

3.   A three minute timer (an egg timer)

4.   A deck of playing cards with all face cards and jokers
     removed. If there are more than 8 players, 2 decks must
     be combined, more than 16 players, 3 decks, etc.

5.   The 9 Letter Alphabet Dictionary in Alphabetic and in
     numerical sequence for each player.

<u>The Play</u>

The cards are shuffled and spread face down on the table. All
players select a card. High card holder is the dealer for the
first hand. The deal rotates one player at a time to the left,
with each player dealing one hand.

The dealer shuffles the cards and the player to the right cuts
the cards. Each player is dealt 5 cards, one at a time, beginning with the player to the left of the dealer and continuing
clockwise.

All cards remain <u>face down</u> until all players have received 5
cards. The dealer sets the timer for 3 minutes and says
"CREATE." The players then all pick up their cards and play
begins.

When the 3 minutes are up, the dealer says " TIME OF REST."
All players must place their cards face down, place pencils
down and stop using their calculator.

The players, starting on the dealer's left, declare their hands
and the number of points claimed. The dealer verifies the
CREATE declared and the player is given the correct (or corrected)
points. Each player, in turn clockwise, declares until the
dealer declares.

The deal is passed to the left and the next hand is dealt.
When, at the end of a hand, one or more players have 200 or
more points, the player with the most points wins. <u>The objective
of CREATE is to make the most points possible using 5 numbers
from 1 to 10, by adding, subtracting, multiplying, and/or
dividing in any sequence desired.</u> But, don't forget, you can
take those game numbers and by using multiples of 10, derive some
interesting alternatives. (e.g. 50, 500, 5000, etc.)

Here are several examples and their scores:

Hand (suit omitted) 3,4,5,7,7, cards are used as

3	+	300	=	300
4	+	40	=	340
5	+	5	=	345
7	+	7000	=	7345
7	+	70000	=	77345

SHELL

Score:	Sum of Digits:	7 + 7 + 3 + 4 + 5		26
	Bonus	All cards used		10
		TOTAL		36

The same hand may be played as

4	+ 40	=	40
7	X 7	=	280
7	÷ 7	=	40
3	+300	=	340
5	+ 5	=	345
			SHE

Score:	Sum of Digits:	3 + 4 + 5	=	12
	Bonus	All cards used		10
		Multiply used		10
		Divide used		10
		TOTAL		42

Here's another way to play the same hand:

3	+ 3	=	3
5	X 5	=	15
4	+ 4	=	19
7	+ 700	=	719
7	+7000		7719
			GILL

Score:	Sum of Digits:	7 + 7 + 1 + 9	=	24
	Bonus	All cards used		10
		Multiply used		10
		TOTAL		44

Here's yet another way:

3	+300,000	=	300000
7	+ 70,000	=	370000
4	+ 4,000	=	374000
5	+ 5,000	=	379000
7	- 7	=	378993
			EGG BLE

Score:	Sum of Digits:	3 + 7 + 8 + 9 + 9 + 3 =	39
	Bonus	All cards	10
			49

The alternatives are almost limitless. Two players with the same
hand will probably declare different points.

CREATE is truly a game of <u>Skill</u> and <u>Chance</u>.

<u>5 Card Draw CREATE Poker</u>

Number of Players: 2 or more

The game of 5 Card Draw CREATE Poker

Each player antes an agreed upon number of poker chips, and
is dealt 5 cards, having a value of 1 through 10.   The
players are given 3 minutes to CREATE a number word and
determine if they wish to use their hand or draw 1 to 5
cards in exchange for a corresponding number of cards from
their hand.

The players then bet on their hand, <u>before</u> drawing cards.
Each player is given the desired number of cards.   The
players are given 3 minutes to CREATE their final number
word.

The players bet on their hand.   The player with the highest
number of points wins all poker chips in the pot.

The equipment required

1.   An electronic calculator for each player

2.   Pencils and paper as for CREATE

3.   A three minute timer.

4.   A deck of playing cards with all face cards and jokers
     removed.   Multiple decks may be used depending on the
     number of players.

5.   The 9 Letter Alphabet Dictionary in alpha/numeric
     sequence for each player.

6.   Poker chips or equivalent.

<u>Dealing the Cards and Play</u>

The cards are shuffled and spread face down.   All players select
a card and high card is the dealer for the first hand.   Each
hand, the deal rotates one player at a time clockwise.

Each player antes an agreed upon number of chips into the pot.
The dealer shuffles the cards.   The player to the right cuts
the cards.   Each player is dealt 5 cards, one at a time,

beginning with the player to the left of the dealer and continuing
clockwise.

All cards remain <u>face down</u> until all players have received 5 cards.
The dealer sets the timer for 3 minutes and says "CREATE". The
players then all pick up their cards and, as in CREATE, determine
the maximum number of points which they can make from their hands.

When the 3 minutes are up, the dealer says "TIME TO BET". The
players then determine if they wish to exchange from 1 to 4 of
their cards for cards still in the deck.

The players, beginning to the left of the dealer, bet on their
hands. The first player places his bet in the pot. The next
player may either pass, and not play   the rest of the hand, or
places an equal amount in the pot, or raise the bet. Betting
continues until all players have either passed or placed an
equal amount in the pot.

Players may continue to CREATE while betting.

The dealer asks the player to his left "HOW MANY CARDS". The
player may ask for 1 to 4 cards. The player places a corresponding
number of cards from his hand into a discard pile and the dealer
gives him the same number of cards from the top of the deck. The
dealer repeats this for each player and himself. The dealer sets
the timer for 3 minutes and says "FINAL CREATE". The players then
determine the final number of points in their hand. When the
3 minutes are up, the dealer says "FINAL BETS". The players then
as before, bet or pass. Betting continues until all players have
either passed or placed an equal amount in the pot.

Players, remaining in the game, place their cards face down. They
declare the number of points in their hands as in CREATE. The
player with the highest number of points wins the pot. In case
of a tie, the pot is split.

<u>7 Card Stud CREATE Poker</u>

Number of players: 2 or more.

The game of 7 Card Stud CREATE Poker

Each player antes an agreed upon number of poker chips, is dealt
two cards face down and one card face up. The players then bet
as before or pass. Betting continues until all players have
either passed or placed an equal amount in the pot.

The dealer then gives each remaining player one card face up.
Players bet as before.

The dealer then gives each remaining player 2 more cards face
up with the players betting after receiving one card and then
after the second card.

The dealer then gives each remaining player 2 more cards face
up with the players betting after receiving one card and then
after the second card.

The dealer gives the remaining players one card face down, sets
the timer to 3 minutes and says "FINAL CREATE". The players
CREATE the highest score with any 5 of their 7 cards.

When the 3 minutes are up, the dealer says "FINAL BETS". The
players bet as before. The players declare their hands and
the highest score wins the pot.

YOUR   FAVORITE Poker game CREATE

Your favorite Poker game may be played as a CREATE Calc-O-Gram
Game by using the CREATE points to determine the winner.

Wild cards may be jokers or selected face cards added to the
deck. Wild cards may be _any_ number selected by each player
and used in CREATE as that number.

THE 9 LETTER ALPHABET
CALC-O-GRAM DICTIONARY

(Alphabetic Sequence)

The 9 Letter Alphabet Calc-O-Gram Dictionary

We have developed Calc-O-Grams using a 7 letter alphabet. Each of these letters is a fairly true representation when its corresponding number is viewed upside down.

There are 2 more letters that are recognizable when their corresponding numbers are viewed upside down. The 2 letters and their corresponding numbers are:

Letter	Number
G	9
Z	2

This gives us a 9 letter alphabet of

Letter	B	E	G	H	I	L	O	S	Z
Number	8	3	9	4	1	7	0	5	2

Here is a dictionary containing 282 words made from the 9 Letter Alphabet, their corresponding numbers and definitions. (Reference is Webster's New Word Dictionary, College Edition).

This dictionary may be used as the final authority for all Calc-O-Gram games. Plurals, made by the addition of S, are omitted and, of course, are acceptable in games.

WORD	NUMBER	DEFINITION
BE	38	To exist.
BEE	338	An insect.
BEG	938	To ask for as charity.
BEL	738	A unit for expressing in logarithms the ratios of power.
BELIE	31738	To lie about.
BELL	7738	A hollow object which rings when struck.
BELLE	37738	A feminine name.
BESIEGE	3931538	To surround with armed forces.
BESS	5538	A feminine name.
BESSIE	315538	A feminine name.
BEZEL	73238	A sloping surface, as the cutting edge of a chisel.
BEZIL	71238	Same as BEZEL.
BI	18	Two, or slang heteral and homosexual.
BIB	818	Cloth, tied under a child's chin at meals.
BIBB	8818	Part of a ship's mast.
BIBLE	37818	The sacred book of Christianity.
BIG	918	A great size.
BIGGISH	4519918	Somewhat big.
BILBO	0.8718	A long iron bar with shackles that slide back and forth on it.
BILE	3718	Greenish fluid secreted by the liver.
BILGE	39718	Lowest point of a ship's inner hull.
BILL	7718	A statement of charges for goods or services.
BILLIE	317718	A feminine name.
BIO	0.18	A combining form meaning life, of living things.
BIOL	7018	Biological, biologist, biology.
BIS	518	Used as a direction to repeat, especially in music.
BISE	3518	A cold north or northeast wind blowing from the Swiss Alps.
BLE	378	Brotherhood of Locomotive Engineers.
BLEB	8378	Blister.
BLESS	55378	To make holy by spoken formula or sign.
BLG	978	Building.
BLISS	55178	Great joy or happiness.
BLOB	8078	A small mass or splash of color.

BLS	578	Bales, barrels.
BOB	808	Pendant, a float on a fishing line, etc.
BOBBLE	378808	Error, mistake.
BOG	908	Wet, spongy ground.
BOGGLE	379908	To be startled.
BOGIE	31908	A hobgoblin.
BOGLE	37908	A hobgoblin.
BOH	408	Bohemia, Bohemian.
BOHOL	70408	An island in the Philippines.
BOIL	7108	To bubble up and vaporize over direct heat.
BOISE	35108	Capital of Idaho.
BOL	708	Bolivia, Bolivian.
BOLE	3708	A variety of easily pulverized reddish clay.
BOLL	7708	A pod of a plant.
BOLO	0.708	A large single edged knife.
BOO	0.08	A sound to express disapproval.
BOOB	8008	A stupid or foolish person.
BOOHOO	0.04008	To weep noisily.
BOOZE	32008	An alcoholic drink.
BOSH	4508	The lower part of the shaft of a blast furnace.
BOSS	5508	The person in authority.
BOZO	0.208	Slang for fellow or guy.
EBB	883	Receding of the tide.
EBLIS	51783	Satan, the Moslem name.
EEL	733	Any of a group of fishes with long snakelike bodies and no pelvic fins.
EGG	993	The oval body laid by a female bird.
EGGSHELL	77345993	The hard brittle covering of a bird's egg.
EGIS	5193	A shield or breast plate used by Zeus.
EGO	0.93	The self.
EH	43	A sound expressing surprise.
EISEGESIS	515393513	Interpretation of a text by reading in one's own ideas.
EL	73	An extension or wing at right angles to the main structure.
ELBE	3873	A river in Bohemia and Germany.
ELEGIZE	3219373	To write an elegy on.
ELI	173	A masculine name.
ELIS	5173	A division of ancient Greece.
ELIZ	2173	Elizabethan.
ELL	773	See EL.

ELLIS	51773	A masculine name.
ELSE	3573	Different, other.
ELSIE	31573	A feminine name.
ESE	353	East by Southeast.
ESS	553	Something shaped like an S.
ESSE	3553	Being, existence, essence.
ESSO	0.553	An Oil Company.
EZ	23	Ezra.
GEB	839	Born.
GEE	339	An exclamation of surprise.
GEESE	35339	Plural of goose.
GEL	739	A jellylike substance.
GELEE	33739	A cosmetic gel.
GEOG	9039	Geographer.
GEOLOGIZE	321907039	To study geology.
GESSO	0.5539	Plaster of Paris.
GHEE	3349	The liquid butter from buffalo milk.
GI	19	A member of the U.S. armed forces.
GIB	819	An adjustable piece of metal or wood for keeping moving parts of a machine in place, a cat.
GIBE	3819	To jeer.
GIE	319	Variation of give.
GIG	919	A light two-wheeled open carriage drawn by one horse.
GIGGLE	379919	To laugh foolishly, as to titter.
GIGOLO	0.70919	A man paid to escort a woman.
GILES	53719	A masculine name.
GILL	7719	Organ for breathing by animals living in water.
GILLIE	317719	A sportsman's attendant or servant.
GILOLO	0.70719	An island in the Netherland Indies.
GLEBE	38379	Land belonging to a church.
GLEE	3379	Gaiety, mirth, joy.
GLEG	9379	Quick in perception.
GLIB	8179	Done in a smooth, offhand fashion.
GLOB	8079	A lump or mass.
GLOBE	38079	Any round, ball-shaped thing.
GLOBOSE	3508079	Globular.
GLOSS	55079	Brightness or luster of a polished surface.
GLOZE	32079	To explain.
GO	0.9	To move.
GOB	809	A lump or mass; a sailor in U.S. Navy.

GOBBLE	378809	To eat quickly.
GOBI	1809	A large desert in Asia.
GOBO	0.809	A dark strip to shield a TV camera from light.
GOES	5309	Third person singular of to go.
GOGGLE	379909	To stare with bulging or wide open eyes.
GOGO	0.909	Dancer or discotheque.
GOO	0.09	Sticky substance.
GOOGOO	0.09009	Loving.
GOOSE	35009	Large water bird.
GOSH	4509	An exclamation.
HE	34	The man, boy or male animal previously mentioned.
HEBE	3834	Greek mythology, goddess of youth.
HEEL	7334	Back part of the human foot, to follow closely.
HEELLESS	55377334	Lacking a heel or heels.
HEIGH	49134	An exclamation.
HEIL	7134	Hail.
HELIO	0.1734	A heliogram.
HELIOS	501734	Greek mythology, the sun god.
HELL	7734	The place where fallen angels, devils live.
HELLO	0.7734	A greeting.
HES	534	He is.
HESSE	35534	A former kingdom and duchy of west central Germany.
HI	14	A greeting.
HIE	314	To strive, hasten, to pant, gasp for air.
HIGH	4914	Lofty, tall.
HILL	7714	A raised part of earth's surface, smaller than a mountain.
HIS	514	That belonging to him.
HISS	5514	To make a sound like a prolonged S.
HOB	804	Play (or raise) hob, to cause mischief.
HOBBLE	378804	To hamper movement by tying two legs together.
HOBO	0.804	A vagrant, tramp.
HOE	304	A garden tool.
HOG	904	A pig.
HOHO	0.404	An exclamation made in laughter.
HOISE	35104	To hoist.
HOLE	3704	A hollowed-out place.
HOLLO	0.7704	A shout or call.
HOS	504	Hosea.
HOSE	3504	A flexible pipe or tube, stocking.

I	1	Self.
IB	81	In the same place.
IBIS	5181	Large wading bird.
IE	31	That is.
IGLOO	0.0791	An eskimo house made of blocks of packed snow.
IHS	541	Iesus Hominum Salvator, Jesus Savior of Men.
ILE	371	Island.
ILL	771	Not healthy.
ILLEGIBLE	378193771	Not readable.
ILO	0.71	International Labor Organization.
ILOILO	0.71071	A seaport in the Philippines.
IO	0.1	Greek mythology, a maiden loved by Zeus.
IS	51	Third person singular, present indicative to be.
ISIS	5151	Egyptian goddess of fertility.
ISL	751	Island.
ISLE	3751	Island.
ISOGLOSS	55079051	Imaginary line between regions differing in some feature of pronunciation.
ISSEI	13551	A Japanese who emigrated to the U.S. after Oriental exclusion proclamation of 1907.
LBS	587	Pounds.
LE	37	Left end.
LEE	337	Side or part away from the wind.
LEES	5337	Dregs, grounds.
LEG	937	Part of body used to stand and walk.
LEGES	53937	Plural of lex (law).
LEGIBLE	3781937	Readable.
LEGIS	51937	Legislature.
LEGLESS	5537937	Without legs.
LEHIGH	491437	A river in eastern Pennsylvania.
LEI	137	Garland or wreath of flowers or leaves.
LEIGH	49137	A masculine name.
LEO	0.37	The fifth sign of the zodiac.
LESBOS	508537	An island in the Aegean.
LESLIE	317537	A feminine name.
LESS	5537	Not so much.
LESSEE	335537	Tenant.
LI	17	A Chinese measure of distance, about one-third of a mile.
LIB	817	Liberal or as in Women's LIB.
LIBEL	73817	Written statement to injure a person's reputation.
LIBELEE	3373817	Defendant in a suit by libel.

LIE	317	To make a false statement.
LIEGE	39317	Bound to give service and allegiance to the lord.
LILLE	37717	A city in northern France.
LISLE	37517	A fine, hard, extra-strong cotton thread.
LLB	877	Bachelor of Laws.
LOB	807	To throw, toss slowly and in a high curve.
LOBE	3807	Fleshy lower end of the human ear.
LOBO	0.807	The larger, gray timber wolf of the western U.S.
LOESS	55307	A fine grained, yellowish brown loam deposited by the wind.
LOG	907	A section of a felled tree.
LOGE	3907	Any of a series of compartments in a theater.
LOGO	0.907	Motto or symbol.
LOGOS	50907	Controlling principle of the universe and as being manifested by speech.
LOIS	5107	A feminine name.
LOLL	7707	To lean or lounge about.
LOO	0.07	A card game.
LOOBIES	5318007	Awkward, clumsy fellows.
LOOIE	31007	Slang for lieutenant.
LOOSE	35007	Not confined, free.
LOSE	3507	To cease to have.
LOSEL	73507	Worthless.
LOSS	5507	A losing or being lost.
OB	80	A river in western Siberia.
OBELIZE	3217380	To make with an obelus, reference mark.
OBESE	35380	Very fat.
OBI	180	A broad sash with a bow in the back worn by Japanese women and children.
OBLIGE	391780	To compel by moral, legal or physical force.
OBLIGEE	3391780	A person obliged to another.
OBOE	3080	A reed musical instrument.
OBOL	7080	An obolus, an ancient Greek coin.
OBS	580	Obsolete.
OBSESS	553580	To haunt or trouble in mind.
OGEE	3390	Any S-shaped curve or line.
OGLE	3790	To keep looking at.
OH	40	An exclamation expressing surprise.
OHIO	0.140	A state of the United States.
OHO	0.40	An exclamation expressing surprise, taunting, etc.

OIL	710	Petroleum.
OISE	3510	A river in Belgium and France.
OLEO	0.370	Oleomargarine.
OLIO	0.170	A highly spiced stew of meat and vegetables.
OOZE	3200	Soft mud or slime.
OS	50	A bone.
OSLO	0.750	The capital of Norway.
OSS	550	Office of Strategic Services.
OZ	20	Ounce.
SB	85	Substantive.
SBE	385	South by East.
SEE	335	To observe visually.
SEEL	7335	To close the eyes.
SEGO	0.935	Bulb plant with trumpet-shaped flowers.
SEI	135	Small white spotted rorqual fish.
SEISE	35135	In law, to take possession of.
SEIZE	32135	To take possession.
SELL	7735	To give up in exchange for money.
SESSILE	3715535	Attached directly by its base.
SEZ	235	Slang for says.
SHE	345	The female object or person previously mentioned.
SHELL	77345	A hard outer covering.
SHES	5345	She is.
SHIEL	73145	Shieling (Scots dialect for shilling).
SHILL	77145	A confederate of a gambler, barker, peddler, etc.
SHILOH	407145	A national military park in southwestern Tennessee.
SHOE	3045	Outer covering for human foot.
SHOO	0.045	An exclamation used for scaring away.
SIB	815	A person's blood relative.
SIEGE	39315	Persistent attempt to gain control.
SIGH	4915	A long, deep, audible breath.
SIGIL	71915	A seal; signet.
SILL	7715	Bottom frame of a door or window.
SILO	0.715	An airtight pit or tower.
SIS	515	Sister.
SIZE	3215	The quantity of a thing which determines how much space it takes up.
SIZZLE	372215	To make a hissing sound when in contact with heat.
SLEIGH	491375	Snow vehicle with runners.
SLIGO	0.9175	A county in northwestern Ireland.
SLOB	8075	Sloppy, stupid, clumsy person.

SLOE	3075	A small, blue-black, plumlike fruit.
SLOG	9075	To make one's way heavily, plod.
SLOSH	45075	To splash or move clumsily through water, mud, etc.
SO	0.5	As stated or described.
SOB	805	To weep aloud.
SOHO	0.405	A French, Italian and Swiss quarter of London.
SOIL	7105	Layer of earth.
SOL	705	The sun.
SOLE	3705	Bottom surface of the foot.
SOLO	0.705	Made or done by one person.
SOO	0.05	Sault Sainte Marie Canal.
SOS	505	Distress signal.
SOSO	0.505	Indifferently.
ZEE	332	The letter Z.
ZIG	912	A sharp alteration or change of direction.
ZOO	0.02	Collection of living animals for public display.
ZOOGEOG	9039002	Abbreviation for zoogeographic.

(Numerical Sequence)

The various Calc-O-Gram games require a quick reference to the corresponding numbers to each word. One must have either a tremendous memory or a list of the words in numerical sequence.

In this dictionary, the corresponding word numbers are shown in sequence and grouped by number of letters in the word.

Numbers, beginning with other than zero, are grouped for words with 1, 2, 3, 4, 5, 6, 7, 8, and 9 letters.

Those numbers beginning with 0 are presented separately and are grouped for words with 2, 3, 4, 5 and 6 letters. These numbers, when used by themselves or to the extreme right on the calculator display, must have a decimal point substituted for the first 0. However, when another word is to the right, on the calculator display, the numbers are used as shown.

This dictionary may be used during play for all Calc-O-Gram games.

Words beginning with 0

2 letter words

01	IO
04	HO
05	SO
09	GO

3 letter words

002	ZOO
005	SOO
007	LOO
008	BOO
009	GOO
018	BIO
037	LEO
040	OHO
071	ILO
093	EGO

4 letter words

0045	SHOO
0140	OHIO
0170	OLIO
0208	BOZO
0370	OLEO
0404	HOHO
0405	SOHO
0505	SOSO
0553	ESSO
0705	SOLO
0708	BOLO
0715	SILO
0750	OSLO
0804	HOBO
0807	LOBO
0809	GOBO
0907	LOGO
0909	GOGO
0935	SEGO

5 letter words

00791	IGLOO
01734	HELIO
05539	GESSO
07704	HOLLO
07734	HELLO
08718	BILBO
09175	SLIGO

6 letter words

004008	BOOHOO
009009	GOOGOO
070719	GILOLO
070919	GIGOLO
071071	ILOILO

Words beginning with other than 0

1 letter words

1	I

2 letter words

14	HI
17	LI
18	BI
19	GI
20	OZ
23	EZ
31	IE
34	HE
37	LE
38	BE
40	OH
43	EH
50	OS
51	IS
73	EL
80	OB
81	IB
85	SB

3 letter words

135	SEI
137	LEI
173	ELI
180	OBI
235	SEZ
304	HOE
314	HIE
317	LIE
319	GIE
332	ZEE
335	SEE
337	LEE
338	BEE
339	GEE
345	SHE
353	ESE
371	ILE
378	BLE
385	SBE
408	BOH
504	HOS
505	SOS
514	HIS
515	SIS
518	BIS
534	HES
541	IHS
550	OSS
553	ESS
578	BLS
580	OBS
587	LBS
705	SOL
708	BOL
710	OIL
733	EEL
738	BEL
739	GEL
751	ISL
771	ILL
773	ELL
804	HOB
805	SOB
807	LOB
808	BOB
809	GOB
815	SIB
817	LIB
818	BIB
819	GIB
839	GEB
877	LLB
883	EBB
904	HOG
907	LOG
908	BOG
912	ZIG
918	BIG
919	GIG
937	LEG
938	BEG
978	BLG
993	EGG

4 letter words

1809	GOBI
2173	ELIZ
3045	SHOE
3075	SLOE
3080	OBOE
3200	OOZE
3215	SIZE
3349	GHEE
3390	OGEE
3504	HOSE
3507	LOSE
3510	OISE
3518	BISE

3553	ESSE		7708	BOLL
3573	ELSE		7714	HILL
			7715	SILL
3704	HOLE		7718	BILL
3705	SOLE		7719	GILL
3708	BOLE		7734	HELL
3718	BILE		7735	SELL
3751	ISLE		7738	BELL
3790	OGLE			
3807	LOBE		8008	BOOB
3819	GIBE		8075	SLOB
3834	HEBE		8078	BLOB
3873	ELBE		8079	GLOB
3907	LOGE		8179	GLIB
			8378	BLEB
4508	BOSH		8818	BIBB
4509	GOSH			
4914	HIGH		9039	GEOG
4915	SIGH		9075	SLOG
			9379	GLEG

5 letter words

5107	LOIS		13551	ISSEI
5151	ISIS			
5173	ELIS		31007	LOOIE
5181	IBIS		31573	ELSIE
5193	EGIS		31738	BELIE
5309	GOES		31908	BOGIE
5337	LEES			
5345	SHES		32008	BOOZE
5507	LOSS		32079	GLOZE
5508	BOSS		32135	SEIZE
5514	HISS			
5537	LESS		33739	GELEE
5538	BESS			
			35007	LOOSE
7718	BILL		35009	GOOSE
7080	OBOL		35104	HOISE
7105	SOIL		35108	BOISE
7108	BOIL		35135	SEISE
7134	HEIL		35339	GEESE
7334	HEEL		35380	OBESE
7335	SEEL		35534	HESSE
7707	LOLL			

37517	LISLE
37717	LILLE
37818	BIBLE
37908	BOGLE
38079	GLOBE
38379	GLEBE
39315	SIEGE
39317	LIEGE
39718	BILGE
45075	SLOSH
49134	HEIGH
49137	LEIGH
50907	LOGOS
51773	ELLIS
51783	EBLIS
51937	LEGIS
53719	GILES
53937	LEGES
55079	GLOSS
55178	BLISS
55307	LOESS
55378	BLESS
70408	BOHOL
71238	BEZIL
71915	SIGIL
73145	SHIEL
73238	BEZEL
73507	LOSEL
73817	LIBEL
77145	SHILL
77345	SHELL

6 letter words

315538	BESSIE
317537	LESLIE
317718	BILLIE
317719	GILLIE
335537	LESSEE
372215	SIZZLE
378804	HOBBLE
378808	BOBBLE
378809	GOBBLE
379908	BOGGLE
379909	GOGGLE
379919	GIGGLE
391780	OBLIGE
407145	SHILOH
491375	SLEIGH
491437	LEHIGH
501734	HELIOS
508537	LESBOS
553580	OBSESS

7 letter words

3217380	OBELIZE
3219373	ELEGIZE
3373817	LIBELEE
3391780	OBLIGEE
3508079	GLOBOSE
3715535	SESSILE
3781937	LEGIBLE
3931538	BESIEGE
4519918	BIGGISH
5318007	LOOBIES

5537937   LEGLESS

9039002   ZOOGEOG

8 letter words
55079051   ISOGLOSS
55377334   HEELLESS
77345993   EGGSHELL

9 letter words
321907039   GEOLOGIZE

378193771   ILLEGIBLE

515393513   EISEGESIS

39	LEI		78	EEL
40	21 LEIS		79	SHELL OIL
41	OBESE		80	HE LOSES
42	SHE SELLS		81	HIS HEEL
43	SHELLS		82	HIS BILL
44	BESSIE		83	HIS ISLE
45	BILLIE		84	IS LESS
46	BOIL		85	HIS SOLE
47	OIL		86	HE LOBS
48	BLESS		87	HIS SHOE
49	BLESSES		88	BOSS
50	BOSS		89	LOBOS
51	BLOB		90	HILL
52	BELIE		91	BEE
53	BEE		92	9 LBS
54	BOB		93	HOB
55	EBB		94	LIB
56	HOLE		95	BIBLE
57	ELI		96	SEE
58	ELSE		97	SHELLS
59	ELSIE		98	OIL
60	HE		99	BLOB
61	HEEL		100	ESSOS
62	102 LBS		101	OIL
63	LEE			
64	ISLE			
65	LESS			
66	HE LIES			
67	OBSESS			
68	OBOE			
69	SEE			
70	LESLIE			
71	SHE			
72	SOS			
73	SOIL			
74	OH HELL			
75	BESS			
76	2 BOLOS			
77	EBLIS			